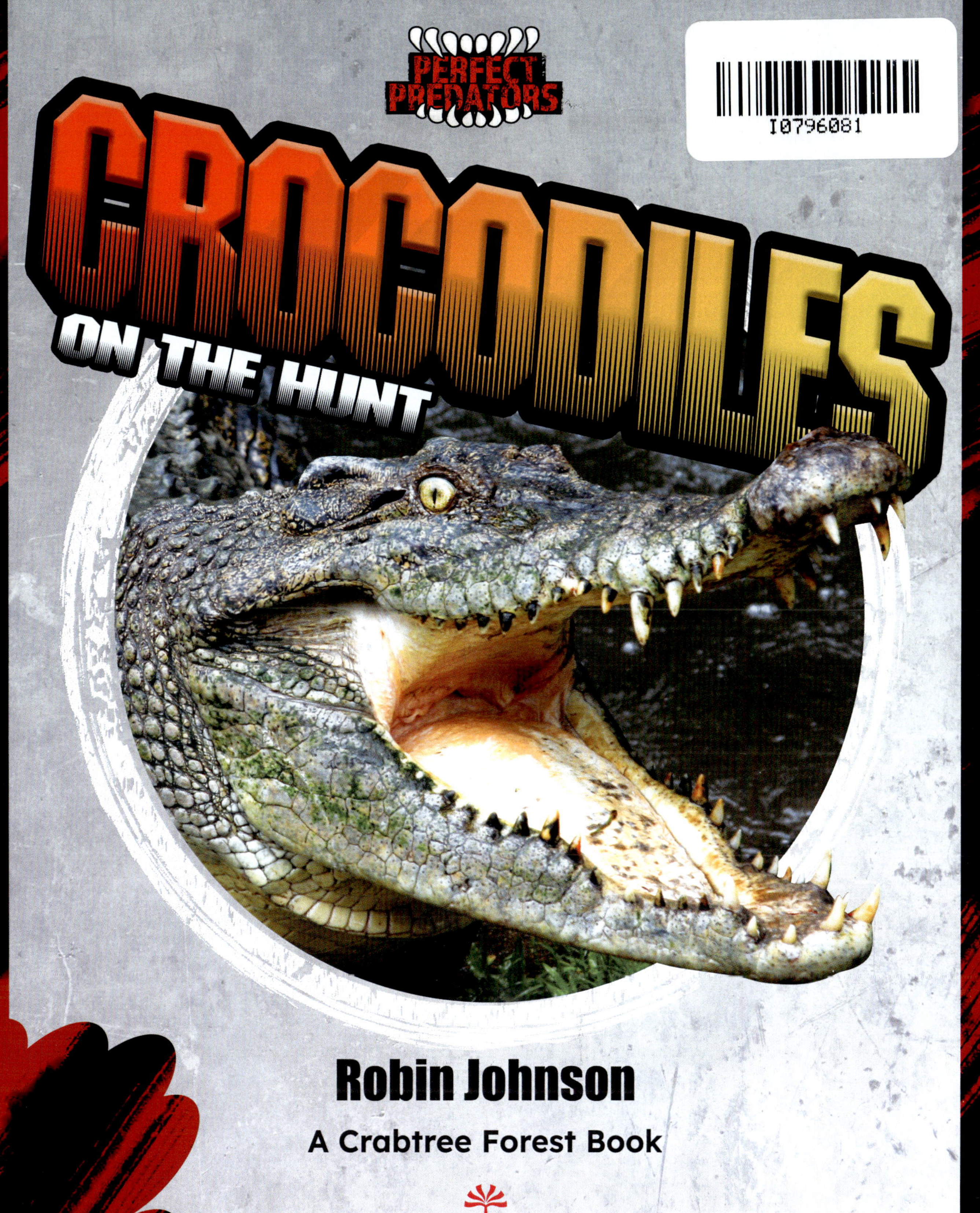

CROCODILES ON THE HUNT

Robin Johnson

A Crabtree Forest Book

Crabtree Publishing
crabtreebooks.com

Notes for Readers and Class Discussion

This book is designed to teach readers about core subject areas, build curiosity, and inspire further investigation. Readers are encouraged to build upon what they already know about the subject and engage in topics they want to learn more about. Here are a few guiding questions to prompt reflection and discussion. Possible answers appear in red.

Before Reading

Read the title and look at the table of contents. What do you already know about predators and crocodiles?

- *I know that predators are animals that hunt and eat other animals.*
- *I know that crocodiles are large predators that hunt prey both on land and in water.*

What would you like to learn about crocodiles?

- *I would like to learn about about how a crocodile ambushes its prey.*
- *I would like to learn how crocodiles compare to alligators.*

During Reading

Pause after reading each page or chapter. What questions do you have about what you read? What are you curious to know?

- *I wonder why crocodiles are found on every continent except Europe and Antarctica.*
- *I'm curious to learn more about how crocodiles may work together to hunt and kill prey.*

Make connections with what you are reading. How is this information similar to something you already know?

- *I know that other animals, such as turtles and lizards, rely on the environment to control their body temperature.*
- *I know that climate change is a serious threat to many animals because rising temperatures cause swamps, wetlands, and other freshwater habitats to dry up.*

After Reading

Recall key details about the book. What was the author trying to teach readers?

- *The author was trying to teach readers that a crocodile's powerful body, sharp senses, and ambush hunting methods make it one of nature's top predators.*
- *The author was trying to teach readers that the biggest threat to crocodiles is the loss of their natural habitats through human actions, leaving crocodiles with fewer places to live and hunt.*

How did the images and captions help you understand more?

- *The diagrams and labels helped show me how a crocodile's different body parts help it find and attack prey.*
- *The photographs showed me how well a crocodile can blend into its surroundings to ambush unsuspecting prey.*

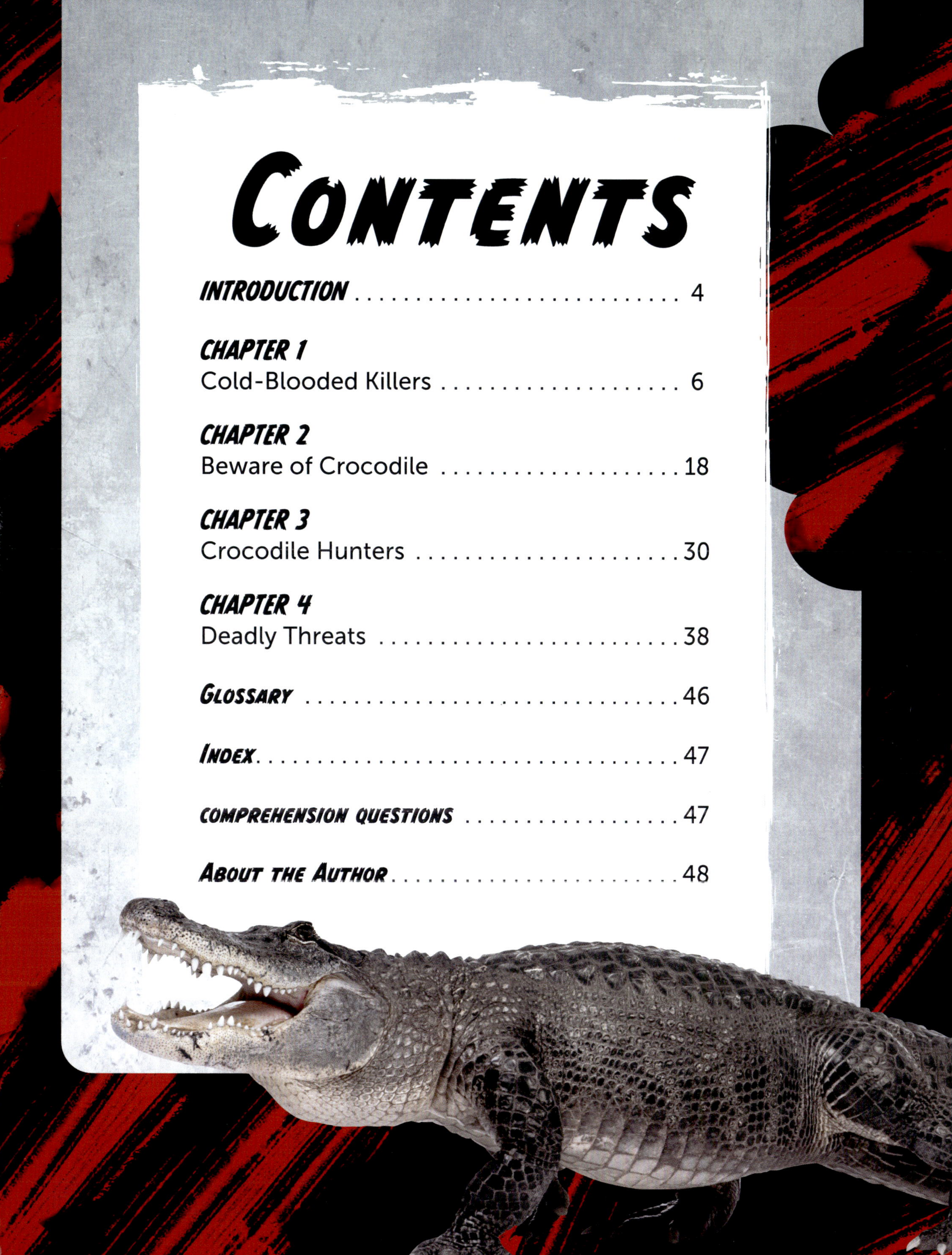

CONTENTS

DROWN... RIP... DEVOUR

A thirsty wildebeest walks cautiously toward a river. It scans the area and sees no signs of danger before it dips its head into the cool water and begins to drink. Suddenly, what looked like a harmless log springs to life—it is actually a mighty crocodile, one of nature's perfect **predators**! With a quick snap of its jaws, the crocodile grabs the antelope and pulls it underwater. The crocodile spins around at great speed, dizzying and drowning its **prey**. Then it rips the wildebeest apart and devours it.

Crocodiles are perfect predators that use the element of surprise to hunt unsuspecting prey.

WHAT IS A PERFECT PREDATOR?

Perfect predators are lurking on land and in waters around the world. These highly skilled hunters are bigger, stronger, and faster than other **carnivores**. With powerful jaws, razor-sharp teeth, and other deadly body parts, perfect predators can catch and kill even large prey animals and defend their spots at the top of the **food chain**.

CHAPTER 1

COLD-BLOODED

KILLERS

Crocodiles are ferocious reptiles that **stalk**, **ambush**, and kill unsuspecting animals. Their big, strong bodies and powerful jaws are built to swiftly and surely capture prey of all kinds.

BIG STUFF!

Think Big

Crocodiles are the largest and heaviest reptiles on Earth. Although crocodile species vary greatly in size, they are all mighty beasts with huge snouts and long, powerful tails. Even the smallest species, the African dwarf crocodile, reaches lengths of 6 feet (1.8 m) from the tip of its snout to the end of its tail. The saltwater crocodile is the biggest species of all and is an absolute monster. It can grow up to 23 feet (7 m) long—about four times the height of an adult human—and weigh a whopping 2,600 pounds (1,200 kg).

Although *Deinosuchus* (see page 9) mostly ate aquatic animals, it also hunted large dinosaurs.

Ruling Reptiles

Dinosaurs were once the ruling reptiles, but today crocodiles hold that title. They are the deadliest reptiles on Earth, larger and more powerful than most animals they meet—and then eat. The largest crocodile species are bigger and badder than alligators, green anacondas, and even dragons (Komodo dragons, that is).

TERROR CROCODILES

Millions of years ago, perfect predators called *Deinosuchus* roamed Earth. These enormous **crocodilians** grew up to 33 feet (10 m) long, weighed as much as fully grown elephants, and had teeth the size of bananas. They preyed on even the largest dinosaurs and lived up to their name—terror crocodiles.

AMERICAN ALLIGATOR
15 feet (4.6 m)

SALTWATER CROCODILE
23 feet (7 m)

KOMODO DRAGON
10 feet (3 m)

GREEN ANACONDA
20 feet (6 m)

BUILT TOUGH

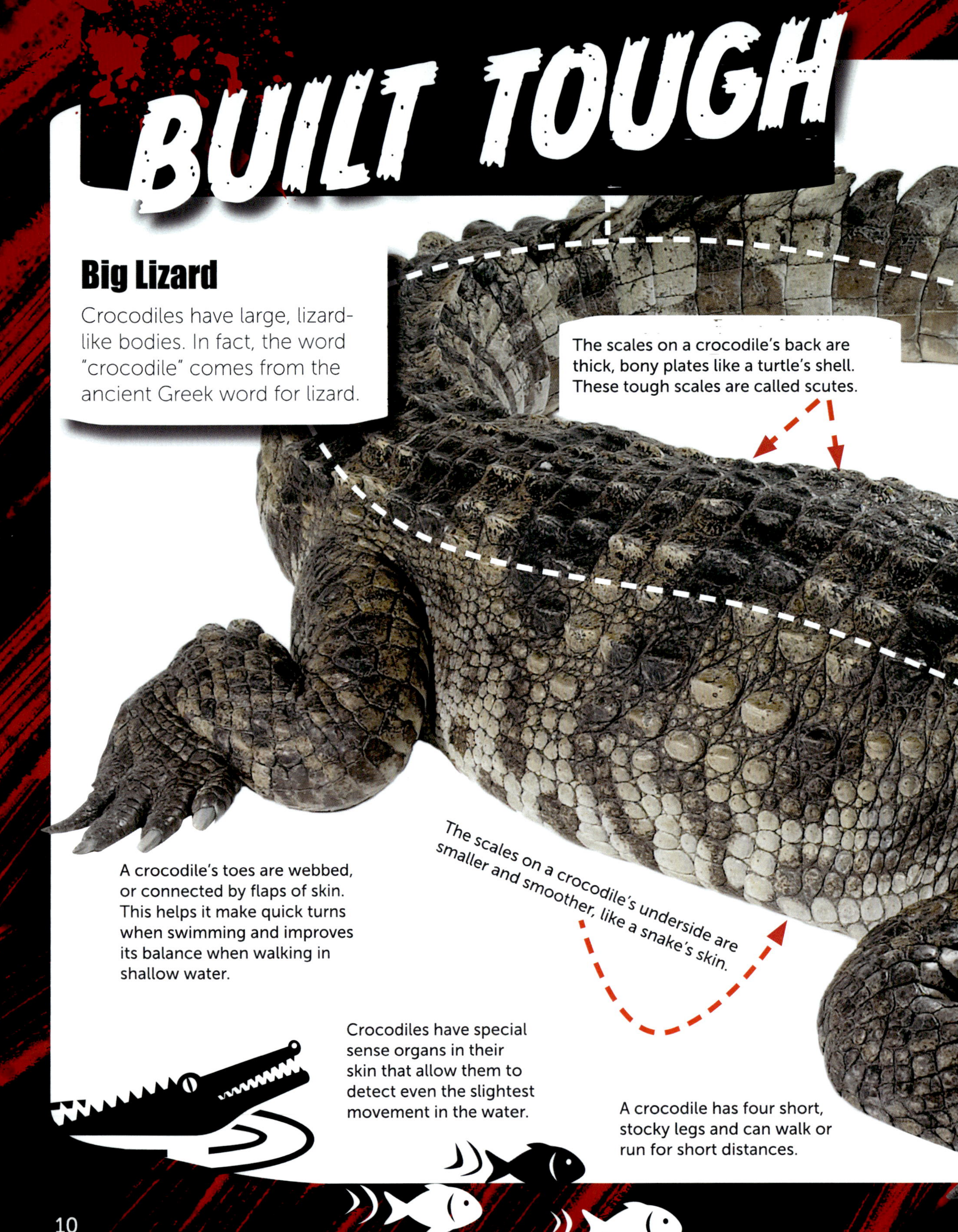

Big Lizard

Crocodiles have large, lizard-like bodies. In fact, the word "crocodile" comes from the ancient Greek word for lizard.

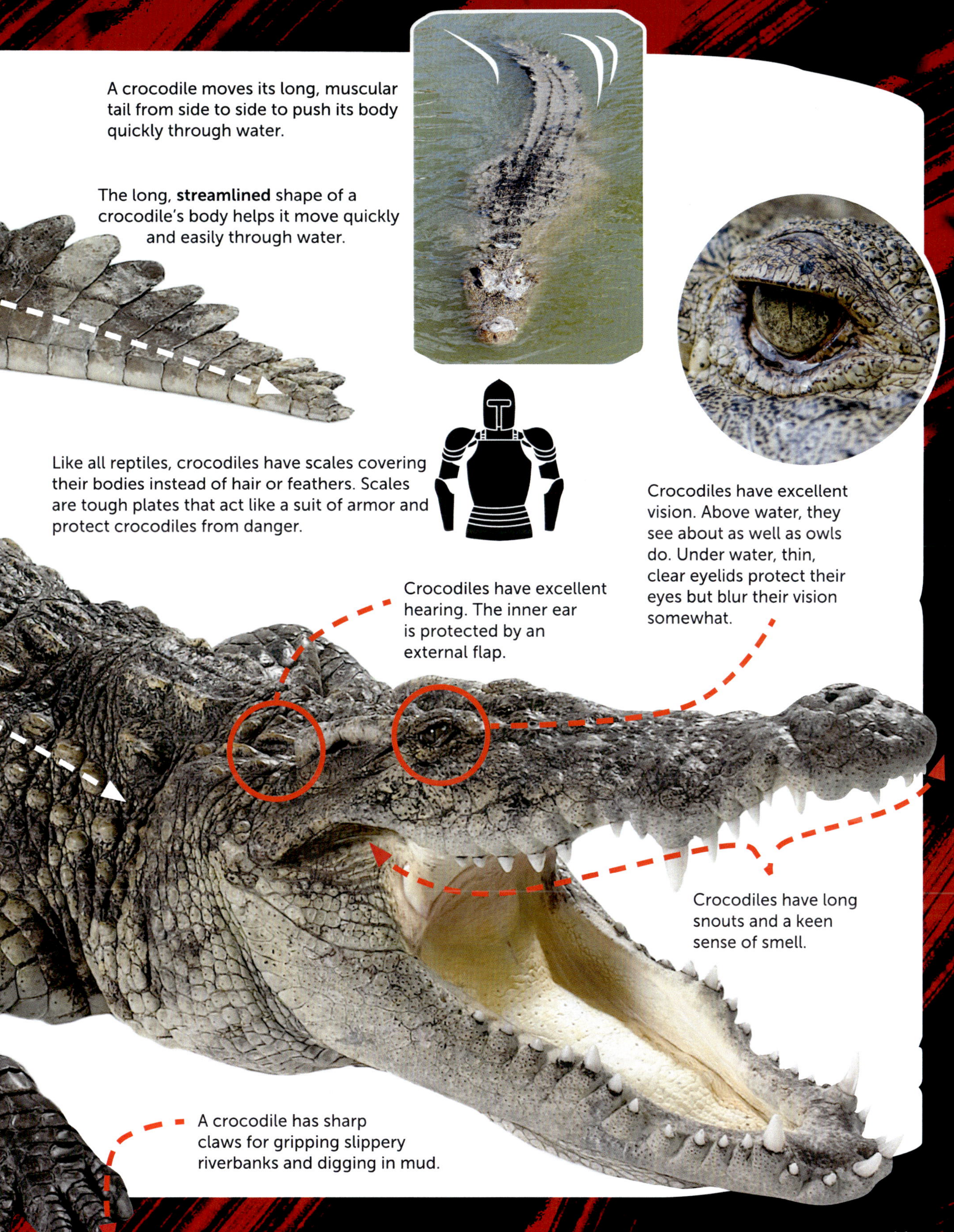
A crocodile moves its long, muscular tail from side to side to push its body quickly through water.
The long, **streamlined** shape of a crocodile's body helps it move quickly and easily through water.
Like all reptiles, crocodiles have scales covering their bodies instead of hair or feathers. Scales are tough plates that act like a suit of armor and protect crocodiles from danger.
Crocodiles have excellent vision. Above water, they see about as well as owls do. Under water, thin, clear eyelids protect their eyes but blur their vision somewhat.
Crocodiles have excellent hearing. The inner ear is protected by an external flap.
Crocodiles have long snouts and a keen sense of smell.
A crocodile has sharp claws for gripping slippery riverbanks and digging in mud.

Jaws of Steel

A crocodile's most terrifying feature is its jaws. They can snap shut very fast and with incredible force. Some crocodile species have more powerful bites than lions, tigers, and bears. Inside a crocodile's jaws are 60 to 100 large, sharp, cone-shaped teeth made for clamping onto animals in a death grip and then tearing them to pieces.

A crocodile's powerful jaws are designed to catch and hold prey. Once an animal has been caught it is very difficult for it to get away.

Gentle Jaws

Female crocodiles are smaller than males, but their jaws are still huge and deadly. Despite their fierce nature, mother crocodiles use their jaws in a surprisingly gentle way—to carry their young. Females bury their eggs in sand along rivers and guard them from predators for up to three months. When the eggs hatch, the mothers dig the babies out of the sand and carefully carry them in their mouths to the water.

A mother crocodile carries her newly hatched baby to water.

PREDATOR POP QUIZ

The animals shown on this page all bite with incredible force—enough to crush a bowling ball. Which one do you think has the strongest bite? Put them in order based on their bite force, with number 1 being the strongest and number 6 being the weakest. Then check your answers below.

Answers: 1-saltwater crocodile, 2-hippopotamus, 3-jaguar, 4-gorilla, 5-spotted hyena, 6-grizzly bear

Meet the Family

There are more than a dozen species of true crocodiles—and they are all uniquely **adapted** to be a perfect predator. A true crocodile is reptile that is part of the crocodile family. Each species has the same main body parts, but with some key differences.

The African dwarf crocodile has a small body, long legs, and a short, upturned snout. It has tough scutes on its underside that help protect it from predators.

Slender-snouted crocodiles are medium-sized crocodiles with very long, slim snouts. They use their snouts to catch fish and other small animals in the water.

The giant saltwater crocodile has fewer scutes on its back than other crocodiles do. Its huge size and powerful jaws give this beast all the protection it needs.

The mugger has the widest snout of all the crocodiles. This species has been observed balancing twigs on its big snout to lure nest-building birds toward it—and then gobbling them up.

The Cuban crocodile spends more time on land than other crocodiles do. It is a medium-sized species with long, strong legs for walking, **galloping**, and jumping.

The Nile crocodile is the second-largest crocodile species. It has a huge, muscular body for hunting and killing large prey animals.

NILE CROCODILE

Nile crocodiles and American alligators are both huge reptiles with long, muscular bodies and powerful jaws. But Nile crocodiles are much bigger and more aggressive than their alligator cousins. In fact, Nile crocodiles are considered the most dangerous crocodilians in the world.

Like most crocodiles, the Nile crocodile has a pointed, V-shaped snout.

Most crocodiles are olive green, light gray, or tan in color.

The Nile crocodile has extremely powerful jaws. It is second only to the saltwater crocodile in bite force.

The Nile crocodile grows up to 20 feet (6 m) long and weighs up to 1,650 pounds (750 kg).

VS ALLIGATOR

Alligators have wider, rounded, U-shaped snouts.

The American alligator has one of the strongest bites of any animal ever measured—almost as powerful as the huge Nile crocodile.

An alligator's top jaw is bigger than its bottom jaw, so its bottom teeth are not visible when its mouth is closed. You can see some of a crocodile's huge bottom teeth when its jaws are clamped shut.

Alligators are usually dark gray or black in color.

The American alligator reaches up to 15 feet (4.5 m) in length and weighs up to 1,000 pounds (450 kg).

CHAPTER 2

BEWARE

OF CROCODILE

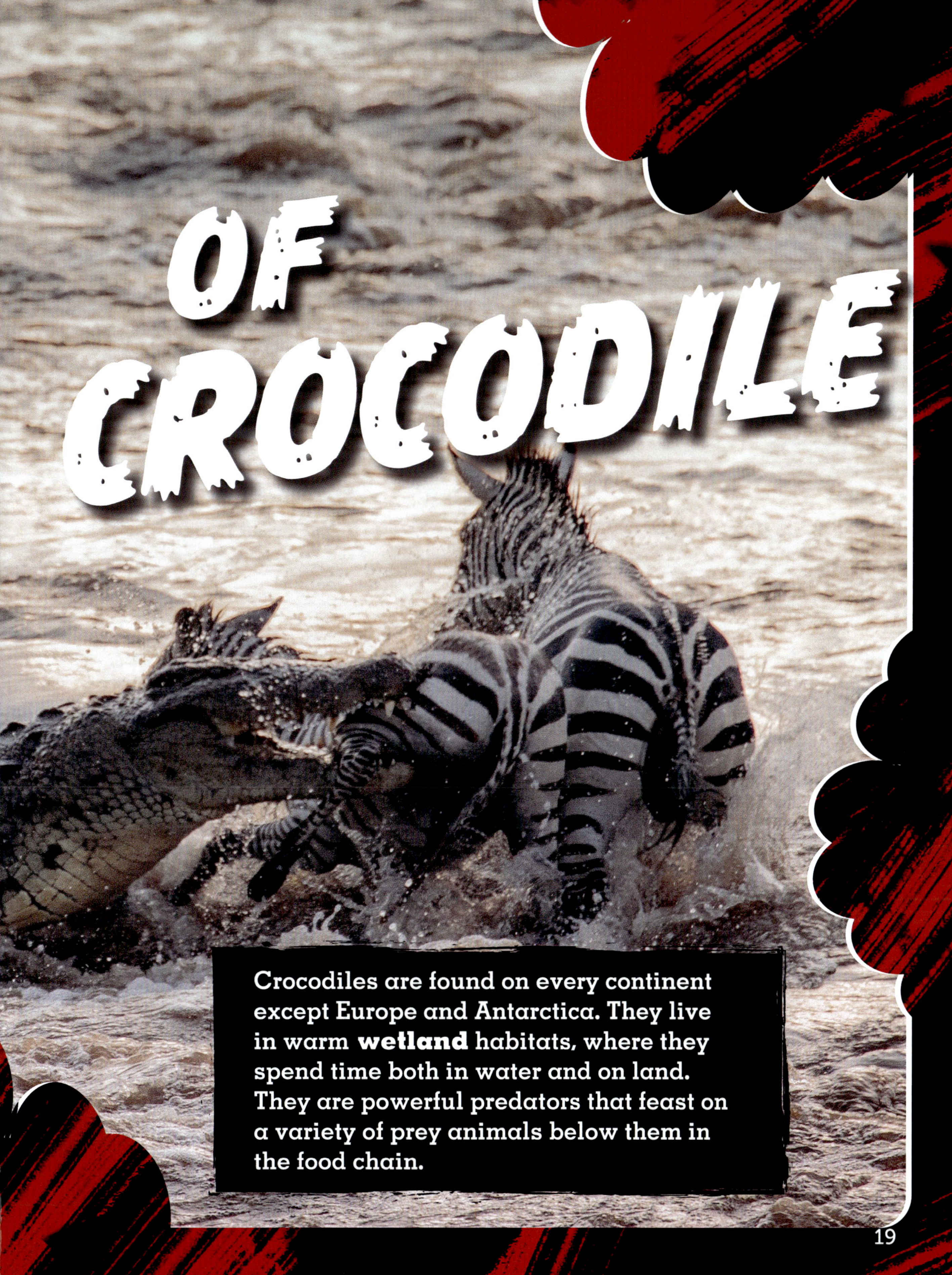

Crocodiles are found on every continent except Europe and Antarctica. They live in warm **wetland** habitats, where they spend time both in water and on land. They are powerful predators that feast on a variety of prey animals below them in the food chain.

GLOBAL THREAT

Miles of Crocodiles

Different crocodile species are found in different parts of the world. Most live in freshwater **swamps**, rivers, and lakes. A few species live mainly in saltwater. They make their homes in **estuaries**, **mangroves**, and other brackish, or slightly salty, habitats. They may also swim long distances in the ocean.

The American crocodile is the only crocodile species found in the United States. It is found in south Florida, where it occasionally meets the American alligator.

Cuban crocodiles are found only in freshwater swamps in Cuba.

Morelet's crocodiles are found in freshwater swamps and marshes in Mexico, Belize, and Guatemala.

The Orinoco crocodile lives in freshwater rivers in Colombia and Venezuela.

The American crocodile is found in the United States, Mexico, and Central and South America. It lives in both saltwater and freshwater habitats, including **marshes**, brackish lakes, and mangrove swamps.

Crocodile range

African dwarf crocodiles are found in rain forests, swamps, and slow-moving rivers in West Africa.

The mugger lives in rivers, lakes, and other freshwater habitats in India and surrounding areas.

The Siamese crocodile is found in Southeast Asia. It makes its home in freshwater habitats, including slow-moving rivers, lakes, and marshes.

Small populations of Philippine crocodiles live in freshwater habitats in the Philippines.

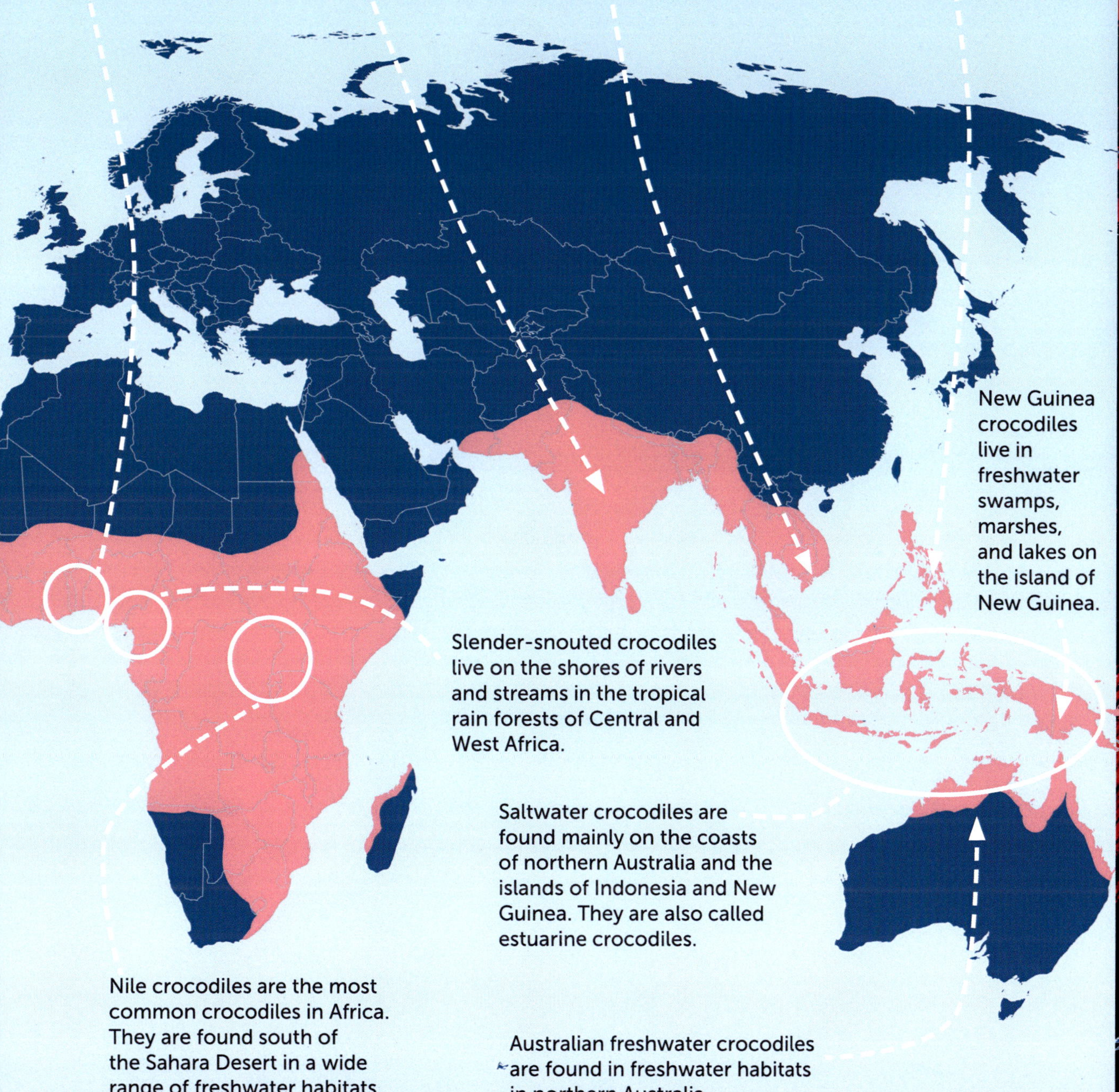

New Guinea crocodiles live in freshwater swamps, marshes, and lakes on the island of New Guinea.

Slender-snouted crocodiles live on the shores of rivers and streams in the tropical rain forests of Central and West Africa.

Saltwater crocodiles are found mainly on the coasts of northern Australia and the islands of Indonesia and New Guinea. They are also called estuarine crocodiles.

Nile crocodiles are the most common crocodiles in Africa. They are found south of the Sahara Desert in a wide range of freshwater habitats, including swamps, lakes, and fast-flowing rivers.

Australian freshwater crocodiles are found in freshwater habitats in northern Australia.

FRESHIE

Freshwater crocodiles and saltwater crocodiles both live in northern Australia, but these species have very different habitats, bodies, diets—and habits of preying on people.

Freshwater crocodiles, or "freshies," live in inland swamps, rivers, and other freshwater habitats.

Freshwater crocodiles grow up to 10 feet (3 m) long and weigh up to 150 pounds (70 kg).

Freshwater crocodiles have the weakest bite of any crocodile species. They are timid, or shy, animals that try to avoid salties and humans. However, they will defend themselves and their nests.

Freshwater crocodiles have longer, narrower snouts and smaller, needle-like teeth. They eat fish, frogs, rats, insects, and other small animals.

VS SALTIE

Saltwater crocodiles, or "salties," live mainly along coasts and in areas of rivers that mix with salty ocean waters. They are very strong swimmers and can be found in the ocean far from shore.

Saltwater crocodiles have broader, thicker snouts and bigger teeth for hunting and eating larger prey. Their diet includes snakes, birds, monkeys, deer, and buffalo, as well as cattle and other **livestock**.

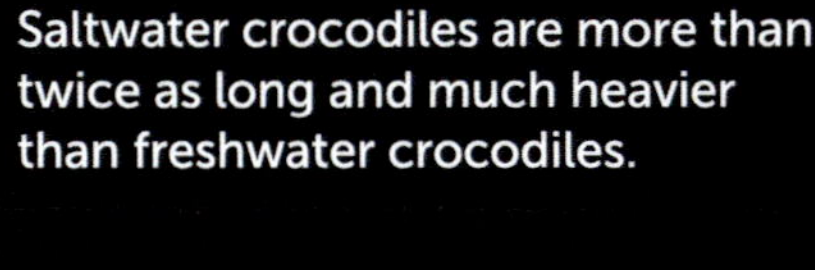

Saltwater crocodiles are more than twice as long and much heavier than freshwater crocodiles.

Saltwater crocodiles are very aggressive and much more likely to attack large animals—including freshies and humans—with their powerful jaws.

COASTAL KILLERS

Float and Bask

Crocodiles are semiaquatic, which means they live and hunt prey both in water and on land. They are excellent swimmers and can float motionless for long periods of time or move swiftly to catch prey in the water. They can also hold their breath for an hour or more while lurking or swimming beneath the surface.

All crocodiles have **glands** in their tongues that remove excess salt and allow them to survive in saltwater habitats for short periods of time.

Warming Up

Crocodiles come out of the cool water to lay in the Sun and warm their bodies. This is called basking. Like all cold-blooded animals, crocodiles rely on their environment to control their body temperature. They lie completely still at the water's edge, soaking up the Sun's rays for hours at a time. But do not be fooled by these sleepy sunbathers. Crocodiles are always alert and can snap to attention—and snap their jaws shut on prey—in an instant.

Crocodiles do not sweat, so they open their mouths to release excess body heat.

DIGGING DEEP

Some crocodiles use their strong, sharp claws to dig dens called burrows along the banks of rivers. They seek shelter in the burrows when the weather is too hot, cold, or dry. Some species also make nests or **estivate** in the burrows.

Some crocodiles dig burrows near rivers to stay cool during the dry season.

CHOW DOWN

Pleased to Eat You

What do crocodiles eat? Anything they want! These **apex predators** will kill and eat just about any animal that comes their way. Smaller crocodiles feed on fish, frogs, birds, crabs, lizards, and other animals. Larger crocodiles also eat bigger prey such as zebras, deer, and wildebeests. Crocodiles are also aggressive **scavengers** that will steal animals killed by other predators.

Crocs on Top

Crocodiles eat many prey animals, but few dare to attack these mighty carnivores. Smaller crocodiles may be hunted by anacondas, jaguars, and even other crocodiles, but they are not their main food source. Very bold—or very hungry—animals may try to kill large crocodiles from time to time, but they usually end up as dinner for the crocodiles.

This crocodile has caught a tasty fish to eat.

Crocodiles often eat dead animals that have washed up in rivers and lakes, such as this dead hippopotamus.

CROCODILE TEARS

Scientists have observed that crocodiles really do seem to shed tears when they eat. But these top predators do not cry because they feel bad about killing their prey. Instead, the tears are likely for cleaning and wetting their dry eyes. The sounds and movements crocodiles make when they tear food apart may also cause the tears to flow.

EVERY LAST BITE

Devouring Entirely

Crocodiles eat every part of the animals they catch—including bones, shells, hooves, and horns. They do not chew their food, so their stomachs must work overtime to **digest** bones and other tough body parts. Crocodiles have extremely strong acid in their stomachs to break down the food. Some crocodiles also swallow stones that help crush and grind food in their stomachs.

POWER OF THE HEART

A crocodile can eat up to 23 percent of its body weight in a single meal. The secret to its eating ability is in its heart. A special **valve** sends blood straight to the crocodile's stomach, helping it produce stomach acid much faster than other animals are able to. This allows it to digest huge amounts of food at once.

Crocodiles have jaws of steel—and stomachs to match.

Check Your Croc Knowledge

PREDATOR POP QUIZ

Test what you've learned about crocs so far and fill in the word that completes the sentence.

1 The tough scales on a crocodile's back are called

____

2 Crocodiles are found on every continent except

______________ and ______________

3 Some crocodiles use their

____

to dig burrows along riverbanks.

4 The

____

crocodile is the largest of all the crocodile species.

5 The word "crocodile" comes from the ancient Greek word for

____

Answers: 1. scutes 2. Europe, Antarctica 3. claws 4. saltwater 5. lizard

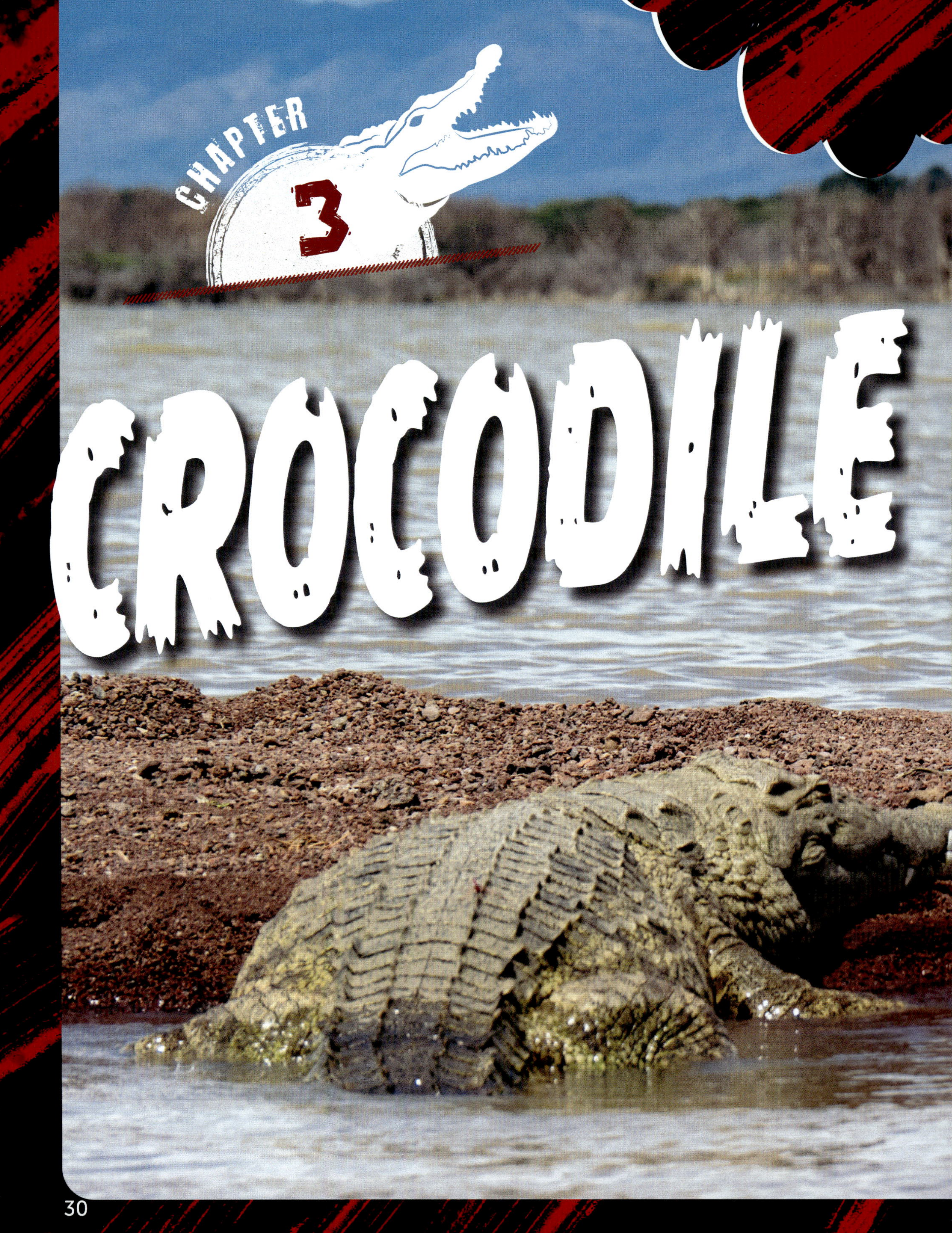

CHAPTER 3

CROCODILE

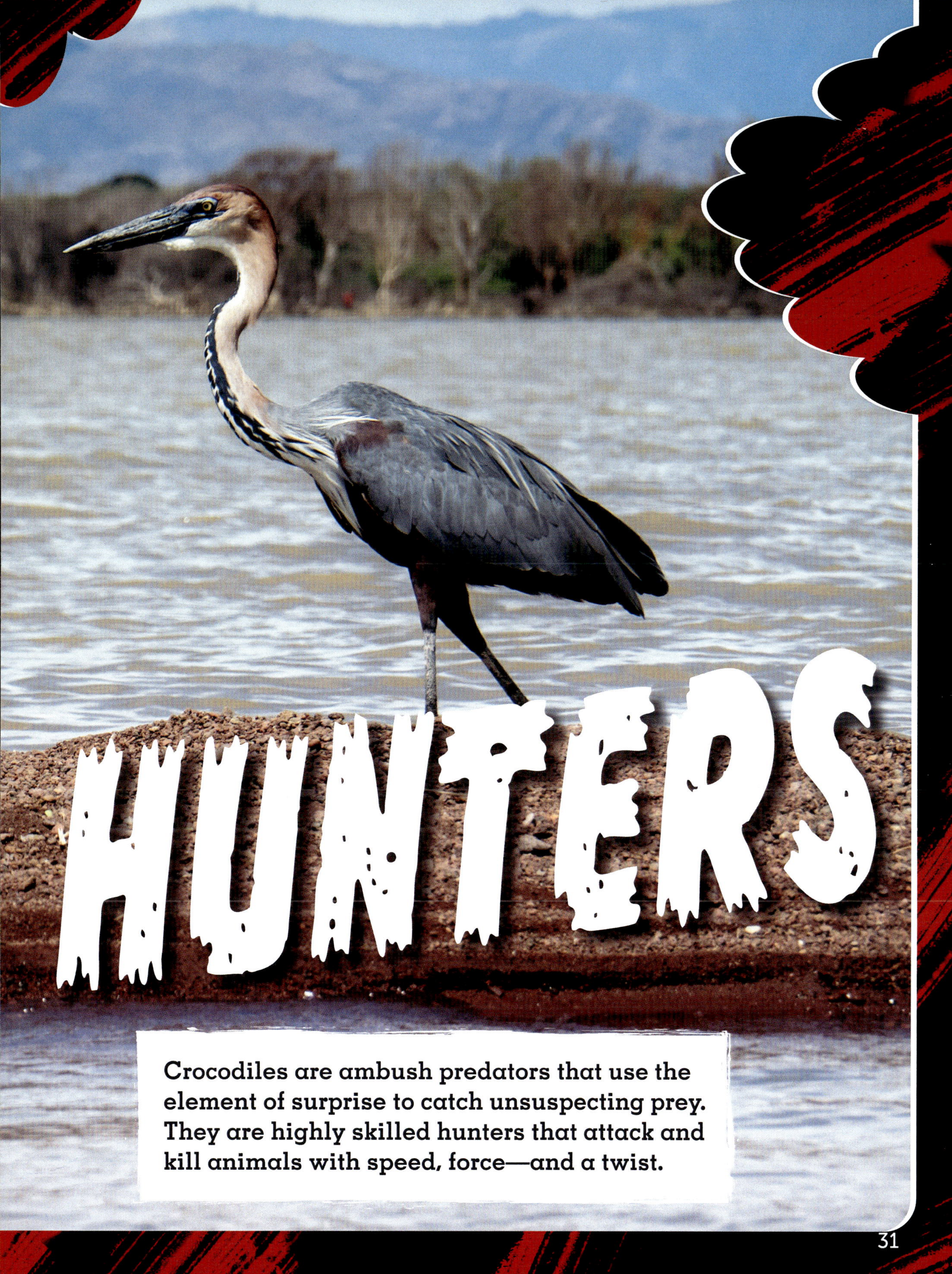

HUNTERS

Crocodiles are ambush predators that use the element of surprise to catch unsuspecting prey. They are highly skilled hunters that attack and kill animals with speed, force—and a twist.

WITHOUT WARNING

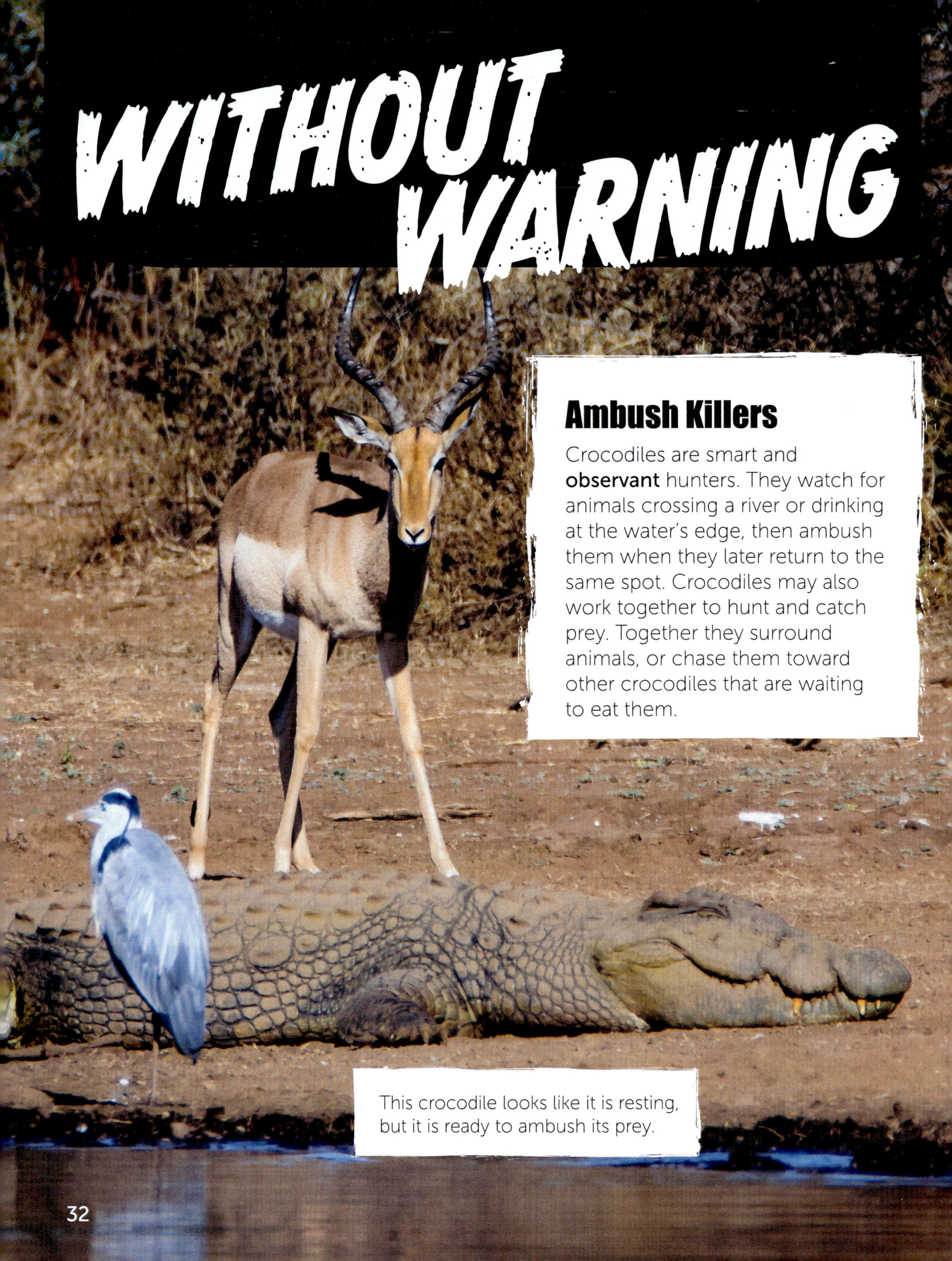

Ambush Killers

Crocodiles are smart and **observant** hunters. They watch for animals crossing a river or drinking at the water's edge, then ambush them when they later return to the same spot. Crocodiles may also work together to hunt and catch prey. Together they surround animals, or chase them toward other crocodiles that are waiting to eat them.

This crocodile looks like it is resting, but it is ready to ambush its prey.